ALSO BY SY MONTGOMERY

Secrets of the Octopus

Of Time and Turtles

The Hawk's Way

The Hummingbirds' Gift

...d Untamed (with Elizabeth Marshall Thomas)

The Soul of an Octopus

Birdology

The Good Good Pig

Search for the Golden Moon Bear

Journey of the Pink Dolphins

Spell of the Tiger

Walking with the Great Apes

The Curious Naturalist

The Wild Out Your Window

...y-three other books, for younger readers,
... Scientists in the Field titles on cheetahs,
..., hyenas, tarantulas, tree kangaroos,
... sharks, snow leopards, and other animals

WH

CH

KI

Tamed ar

And twer
includin
tapi
great whit

WHAT THE CHICKEN KNOWS

A New Appreciation of the World's Most Familiar Bird

SY MONTGOMERY

ATRIA BOOKS

New York   London   Toronto   Sydney   New Delhi

An Imprint of Simon & Schuster, LLC
1230 Avenue of the Americas
New York, NY 10020

An earlier version of this material appeared as a chapter in Sy Montgomery's book *Birdology* (2010).

"Birds Are Individuals" from *Birdology* copyright © 2010 by Sy Montgomery

Introduction copyright © 2024 by Sy Montgomery

Insert photography © Tianne Strombeck

First Atria Books hardcover edition November 2024

ATRIA BOOKS and colophon are trademarks of Simon & Schuster, LLC

Simon & Schuster: Celebrating 100 Years of Publishing in 2024

For information about special discounts for bulk purchases, please contact Simon & Schuster Special Sales at 1-866-506-1949 or business@simonandschuster.com.

The Simon & Schuster Speakers Bureau can bring authors to your live event. For more information or to book an event, contact the Simon & Schuster Speakers Bureau at 1-866-248-3049 or visit our website at www.simonspeakers.com.

Interior design by Jill Putorti

Manufactured in the United States of America

1 3 5 7 9 10 8 6 4 2

Library of Congress Cataloging-in-Publication Data is available.

ISBN 978-1-6680-4736-1
ISBN 978-1-6680-4737-8 (ebook)

For Gretchen Vogel Morin,
who gave us our first flock

INTRODUCTION

Imagine: You are lying on the ground, once again attempting to fix the ancient riding lawn mower you inherited from your home's previous owner. Just beyond the narrow space between the grass and the machine's metal undercarriage, movement catches your eye. Scaly, reptilian, yellow feet, heavily armed with long, pointed claws—and, just above the toes, curved, knife-sharp spurs—are purposefully striding toward you at face level. It feels like that scene from *Jurassic Park*, where the humans are hiding from the hunting velociraptor—an animal from which, in fact, your avian assailant is descended.

You have been detected. The ominous feet hurry now, pounding the ground in a frenzy. There is no mistaking it: this is a rooster on a rampage—the bane of many an otherwise peaceful barnyard.

Unless you're on the receiving end of those spurs, it may look, at least to others, hilarious: a two-foot-tall bird, mostly made of air, weighing only six pounds, can easily scare a 150-pound, six-foot-tall human into running away. But an angry rooster can draw blood. What should you do?

Years ago, my husband joined the ranks of the many backyard chicken farmers, their kids, and their visitors, who have been unexpectedly confronted with this very dilemma. His quite reasonable reaction was, like most of our poultry-keeping colleagues, to spring to his feet and hastily get out of the way.

But now, many years later, my neighbor, Ashley Naglie, suggests a different course of action—one that is staggeringly counterintuitive. In this situation, "It's best," she insists, "to *pick them up*."

What?

"Cuddle them," she continues. "Talk to them in a soft voice. Do your chores while you're holding them close." Watch out for the beak, she advises, because he might bite. If possible, wrap the rooster's feet in a towel or a blanket to protect yourself from his spurs. But gathering the angry bird into your arms and carrying him around is, she assures, the best and only foolproof way to train a rowdy rooster to become a loving friend.

By the time I spoke with Ashley, my husband and I had lived among chickens for decades, including the occasional rooster or two. The guys had all started out as sweet baby chicks, turned into affectionate and assertive young cockerels—and some of them remained perfect gentlemen their whole lives. But occasionally we'd get one who changed. Then what?

We'd never heard of Ashley's training method. When we began keeping chickens in the late 1980s—way before the internet, with its advice for all matters at your fingertips, way before the pandemic craze for backyard poultry—we had no idea how to contend with a rooster gone rogue other than to find him another home. Now we know what to do.

Our travels in what I call the Chicken Universe, as you shall see in the following pages, have been a multi-decade voyage, a journey of revelation. New discoveries continue to this day.

In 2023, a study published in the journal *PLOS One* reported that roosters recognize their own reflections in mirrors—a common (though contested) measure of self-awareness. Previous experiments show that apes, like humans, do this: when chimps who are familiar with mirrors are knocked out and their faces marked with red dye, they'll examine their faces in the glass and touch the red marks, showing they recognize the reflection as their own.

But male birds are famous for attacking their own reflection in the window (especially disturbing when the bird is a cassowary, a five-foot-tall, 120-pound flightless native of Australia and New Guinea, who can break a plate-glass window with a single kick).

Up till last year, roosters failed the mirror test. But then a German scientist tried another way. Sonja Hillemacher, a University of Bonn researcher, gave roosters time in a mirrored enclosure. But how to tell if they recognized themselves? Knowing that roosters warn others of danger, she projected a hawk silhouette over the test roosters to see what they'd do. When another rooster was visible through the partition, the subject of the experiment tried to warn his comrade of the danger. But if there was nothing but mirrors around—no other chickens than himself to be seen—the test rooster was always silent.

The researcher actually found this unsurprising. She knew roosters were smart. The difficult part of her work was devising an experiment that was biologically relevant— one that related to things that the rooster encounters in his daily life. That takes patience and careful observation— exactly what makes Ashley's barnyard findings as eye-opening as those from Hillemacher's lab.

"It takes time and effort to have a gentle rooster," Ashley explains as fat snowflakes fall into her long brown hair.

She and her husband and two kids offer to educate me on this cold January day. We huddle in our parkas inside one of their many spacious backyard coops. Daughter Brooke, fourteen, scoops up a handsome red cockerel named Charlie, who instantly snugs his comb and wattles into her hair, which is long and dark like her mother's. ("They love snuggling into hair," Ashley's husband, Brian, notes.) "Some people grow up thinking roosters are mean," Brooke says. "But you just have to build trust."

"Roosters look big and intimidating," says Brooke's brother, Tyler, age twelve. "But if you treat them right, with respect and kindness, they'll be calm."

This family speaks from experience. They have been rescuing and rehabilitating rowdy roosters for a decade now.

Their travels in the Chicken Universe started when Brooke was five and Tyler was three. While Brian was working as a mechanic, Ashley was then a stay-at-home mom. They purchased two ducks and added six young New Hampshire Red pullets (what hens are called before age one) for the eggs.

The chickens loved playing with the kids. Brooke would often take a youngster on her lap when she played on the swing. The kids napped with the chickens, found worms for them, stroked them. The little flock ranged freely over their lawn and fields while Ashley watched over them.

But one day, during the few minutes when she stepped inside to get Tyler a drink, a fox killed one of the hens. It was Star, one of their favorites. "Star was only three months old. It was devastating. We were all heartbroken." But Ashley knew what she had to do to protect the others. "I realized," she said, "that I needed a rooster."

A rooster watches for danger. With his loud voice, he alerts his hens to any coming threat. And while they take cover, he will often courageously attack the intruder. Because of his spurs and his spunk, a brave rooster may in fact successfully drive away a much larger predator. (Brian, in fact, once watched in amazement as Charlie chased a fox out of their yard.)

Luckily, Ashley's mom, who lives in Vermont, had just brought three roosters over to Brian's grandmother. "They were Easter chicks who turned out to be roosters," Ashley said. Two roosters remained on her grandmother's farm, and Ashley took one, a large and stately golden Buff Orpington whom she named Cooter.

Cooter charmed the family, both avian and human. He was curious and unusually observant. "He loved to watch me work on the car," Brian says. "He'd watch me for five or six hours at a time!" Then, when Brian was finished, Cooter (who has since died of old age) would come and intently inspect the shiny tools, as if trying to figure out how to replicate what the mechanic had just accomplished.

With his flock, Cooter "was very proactive, but also kind and gentle," Ashley tells me. "He'd show the hens where the food was—but wouldn't eat it himself, letting them have first dibs. And he always had his eye on the sky, watching for a hawk. If he saw one, he'd scream and I'd come running." The kids adored him and were always carrying him around, snuggling him. He never attacked. He seemed to enjoy their gentle touch.

Cooter's affectionate attention to his flock soon had another effect as well: fertile eggs—resulting in more roosters. Each was possessed of a distinctive personality. Carlos, for instance, was petrified of his father—but from the moment he hatched, he loved cuddling so much he began to follow people around, asking to be picked up. Ashley wrote and self-published a picture book about this legendary snuggler, titled *Carlos Loves Hugs.* And he's smart: he has figured out how the doorknob works, Ashley reports, and lets himself inside the house. He brings Ashley thoughtful gifts. Once, she says, somehow "he unscrewed the plumbing to the soapstone sink—and brought me the ring and the gasket."

Henry, another rooster, hatched out in Ashley's hands. "He spent a lot of time inside," she explains. "He would run to the top of the stairs each morning to crow. He had a favorite teacup he liked to drink from."

And soon there were others. Word of mouth travels fast in the small towns of southwest New Hampshire. People learned about this rooster paradise—and started bringing their unwanted roosters to the doorstep. "People would say, 'I have this rooster, and he's so mean I don't want him,'" Ashley says. The family always took them in.

There was Hey Hey ("He was so mean at first!") and his brother, Toe-Gone (who lost a toe fighting with another rooster). A woman from a neighboring town brought in three roosters at once: Abraham—who likes to crow while sitting in Brian's lap—George, and John. Neighbors down the street brought in a beautiful red-and-black cockerel. They'd been taking an evening walk and heard someone clucking. They knew no chicken could last long alone in the woods. Apparently the owner had just dumped the bird on the side of the road and driven off.

This sort of thing happens all too often. And lately, it's worse. At Amazing Grace Animal Sanctuary in the nearby town of Sullivan, founder Donna Watterson has seen "a wicked increase" in both abandoned chickens and abandoned roosters since the pandemic. "Everybody got chickens," she reports, "but now they're back to school and work and they just get rid of them." The sanctuary is already full of unwanted donkeys, alpacas, goats, mini and standard horses, sheep, and pigs. Now they have taken in twenty chickens and two roosters, Croakie and Pigeon. They were wandering on the side of the road, where a former owner had abandoned them, and were both rescued by separate Samaritans.

Modern Farmer magazine reported in 2023 that farm rescues and sanctuaries around the country have seen huge influxes of chickens and roosters since the pandemic. The problem is overseas, too: as the pandemic waned, Germany's largest animal shelter reported trying to accommodate three times as many hens and roosters as in a normal year, while Britain's RSPCA was struggling to cope with a record 1,562 unwanted chickens.

Why do people consider roosters in particular such a problem? Not all of them attack. But almost all of them crow (at her animal sanctuary, Watterson got a call about one who was blind and crowed twenty-four hours a day), and for this reason, some towns have ordinances against them. Too many amorous roosters prove a bother to their hens. Some ladies develop bald patches from too many guys trying to jump on their backs all the time.

Most people who adopt chicks (often at Easter) don't realize they won't all grow up to be hens. For most chicks, males and females look alike. (There are exceptions, like

Black Sex-Links—a breed created by crossing a Barred Rock hen and a Rhode Island Red cock. The boys have a white dot on the head, while the girls are pure black.) And there are hatcheries where experts examine the sex organs of the birds they ship—no job for an amateur, as they're hidden inside the chicken's body—to guarantee males or females. But "nobody educates people before they take chicks home," laments Watterson.

Another problem: if there are too many roosters in a flock, they may fight each other over access to the hens. For this reason, the Naglies have built five separate enclosures to accommodate their roosters and their hens. Abraham rules his own flock of five hens inside the Fluffy Butt Hutt, a three-by-four-by-five-foot, three-story affair surrounded with a fenced twelve-foot-by-eight-foot run. Charlie sleeps with a turkey lady friend in a former playhouse. Some roosters are housed in bachelor-only quarters (if they can't see any hens, they usually won't fight). Toe-Gone and Hey Hey lived together happily, until Hey Hey was eaten by a bear.

On cold days like the day of my visit, bachelor roosters may spend the night, and sometimes the day, inside the house with the family. At the moment, Charlie's tall red comb is showing signs of potential frostbite. Brooke continues to cradle him while Tyler runs inside to get some ointment.

After an article about the Naglies' rescue farm appeared in the local paper, Ashley started fielding rooster-related calls every day. Her advice:

1. "Sometimes a rooster who seems to be about to attack is actually showing you he likes you. If a rooster seems to be dancing around you, he is trying to see if you belong to him. Pick him up."
2. "If your rooster brings you rocks and shiny objects, as they sometimes do, pick him up. I always pick up my rooster after accepting these gifts."
3. "Never kick a rooster." (Even raising your foot toward him, without actually kicking him, counts.) "Pick him up instead."

At one point, the family was caring for ten roosters in the rescue. Combined with caring for their hens, a flock of ducks, and Brian's beehives, plus managing the Airbnb they now run, and with Ashley's new job (she now works as a substitute teacher), it proved a bit overwhelming. Today, Ashley can't accept any more roosters, but she is happy to help educate people. And she's honored to be caring for all the roosters currently in residence.

"Why did I do this?" Ashley asks. "I did it because I wanted to help *all* of the roosters I could help." She remembers the day one of her ducks, Pow, got sick. "She was egg-bound [a painful and potentially deadly condition when an egg is stuck inside]. I saw her looking to me for help." Ashley gave Pow a warm sitz bath so she could pass her egg safely. "And I knew I had to give every bird a chance."

My new friend's calling reflects the words on a plaque that hangs over the door to our own chicken coop. It was a gift from our next-door neighbor, Jarvis Coffin, about whom you'll soon read, who loved our chickens dearly.

The words are attributed to the Italian mystic and Catholic friar now known as St. Francis of Assisi. "Not to hurt our humble brethren is our first duty to them," he preached, "but to stop there is not enough. We have a higher mission—to be of service to them whenever they require it."

Our chicken coop is empty, but the saint's words are not. Even though no chickens are living with us right now (you'll find out what happened in the pages ahead), these elegant, intelligent, supremely social birds are still an important part of our neighborhood and our lives. I shall always be grateful to them.

My travels in the Chicken Universe have been a portal to an unknown kingdom. All of us see birds every day, and chickens are among the commonest birds we know. Yet again and again, as I watch the hens and roosters in my life,

I am reminded how movingly like us birds can be—and how thrillingly different. This book is my small effort to be of service to them: to enhance our wonder and deepen our respect and our compassion for these common creatures we all think we know. The longer I watch them, the more clearly I see how rich and varied their lives are, as fraught and joyous and changeable as our own.

WHAT THE CHICKEN KNOWS

Howard Mansfield

"Hello, Ladies!"

Even if there is no one in sight, I call out to them whenever I round the corner of the woodpile to enter the barnyard. For even if they're scattered over several acres of lawn and woods and brush—some hunting in the compost pile, others patrolling the neighbor's blueberry patch, some scratching in the leaves by the stone wall—I know they'll come running. A dozen foot-tall, black and black-and-white figures, holding their wings out like tots spreading their arms to keep balance or beating their wings to propel themselves even faster, come racing toward me on scaly, four-toed feet, showing the wild enthusiasm of fans catching sight of a rock star. It's a welcome that makes me feel as popular as the Beatles—even if my personal fan club is composed entirely of poultry.

At times, I suppose, I am less a celebrity than the moral equivalent of the neighborhood ice cream truck. For often, I come bearing food—vegetable peelings from the house, the trimmings from pie crust dough, and sometimes an entire tub of fresh cottage cheese that I buy just for them. My hens, like many pets, particularly enjoy being fed by hand. The lead chicken, standing before me front and center, tilts her head to examine my offering with one skeptical orange eye. Then she seizes the first morsel in her hard black-and-amber beak—and the crowd goes wild. Everyone pecks with great enthusiasm, hard enough to hurt my palm. If, among the buffet, there is one particularly big treat to be had—a single apple core, a baked squash skin—at some point, somebody will seize this. The victor will run some distance, chased by her sisters, until the prize is either stolen or swallowed. This is usually good for about ten minutes of entertainment.

But often, I don't come bearing food. I come just for a visit. I relish these encounters even more. The Ladies don't seem disappointed at all. They mill at my feet, cheerful and excited, for they know I have a different treat in store. They are waiting for me to pick them up, stroke them, and sometimes—yes, I admit it, despite medical warnings of the slight chance of contracting salmonella—kiss their warm, red, rubbery combs.

They also like me to run my hand along the sleek length of their backs. Each will squat, wings slightly raised, neck feathers erected, welcoming my caress. I start at the back of the neck, and when my hand has completed half its journey, the hen will arch her back. I gently close my fingers around her tail feathers until my stroke swoops into the air—rather like the way you would stroke a cat. Then it starts all over again, until the hen has had enough and has reached what we call "overpet." She fluffs her feathers, shakes, and, forti-fied by affection, strolls off to continue her chicken day.

When I crouch to pet one hen, another one might hop up to perch on my thigh, patiently waiting her turn. I talk to them. "Hello, Ladies! How are my Ladies? Did you find good worms today? What was in the compost?" They keep up their end of the conversation with their lilting chicken voices.

Visitors who witness this for the first time are amazed. "I've never seen anything like that!" they say. "I always thought chickens were stupid! Is it possible," they wonder, "that they actually know you?"

Of course they know me. They know the neighbors, too. In the more than two decades that I've been living with chickens, they have formed deep bonds with some—and not with others. Certain individual chickens adored our pig, Christopher Hogwood, who lived for fourteen years in the pen next to the coop in the barn. Some even chose to roost with him, perched atop his great prone bulk, instead of spending the night with flock-mates. But none of the hens has ever bonded with our border collies. They never visit the neighbors across the street—but they adore the retired couple next door. When the Ladies hear Bobbie and Jarvis Coffin's screen door slam, the hens hop over the low stone wall separating our yards and rush to greet them. When Jarvis relaxes in the backyard hammock on summer days, they gather beneath him, and some leap into the air, attempting to join him in his day roost. (So far they haven't succeeded.) The hens mob the couple whenever they try to enter their cars. Usually Bobbie and Jarvis get them some cracked corn, which they keep in their shed just for our hens, and make their getaway while the birds are eating. Sometimes, Bobbie confesses, when she's in a hurry, she sneaks out the door and tiptoes to the car, to avoid a protracted visit with our chickens.

Sometimes, the Ladies don't wait for the Coffins to make the first move. A few of the bolder hens have been known to mount the flight of wooden steps leading to Bobbie and Jarvis's second-story back porch—quite a feat considering the birds are only twice as tall as the steps are deep. They gather, softly discussing their plans, outside the porch door, looking in through the glass panes, trying to catch the attention of their human friends and entice them to come out and play.

Jarvis Coffin with the Ladies

Occasionally the hens come over while the Coffins are hosting a gathering. Their guests are invariably impressed. "I didn't know you had chickens!" people exclaim—and then seem dumbfounded that, like themselves, our chickens simply enjoy visiting their lively, kind neighbors.

Folks use words like "astonishing" to describe such friendships between people and poultry. What's more astonishing, though, is not that these birds know so much about their human neighbors, but that we humans know so little about our neighborhood birds—even one as common and readily recognized as a chicken.

People think they know lots about chickens, and you'd think they would: there are four living chickens for every living person in the world—and since by definition they are domestic fowl (they are a separate subspecies from their wild ancestor, the red jungle fowl of Southeast Asia), all of them live among humans. There are at least four places in the United States named after chickens, including the towns of Chicken, Alaska (which was supposed to be named after a local grouse, the ptarmigan, but nobody could spell it), and Chickentown, Pennsylvania (named after its many poultry farms); in 1939, the Delaware state legislature, braving the opposition of the State Federation of Women's Clubs of Delaware (which wanted the cardinal instead), selected the Blue Hen chicken, a local variety but not a true breed, as its state bird.

Still, chickens are rarely celebrated in our culture and

rarely given the respect they deserve. I once sat next to a man on an airplane who detailed for me at length the attributes of the species: they are stupid, disgusting, filthy, cowardly, occasionally cannibalistic automatons, he said. How had he acquired this opinion? It turned out he had worked at a factory farm—the sort of place where most chickens are raised for food in the United States—in a dirty, overcrowded warehouse that resembled a prison camp.

This is not the best place to get to know someone. Nor is a dinner plate. Yet for most of us, our relationship with chickens is generally of a culinary nature. In fact, the first definition for the word "chicken" I encountered on the internet doesn't even mention that it's a bird. It's "the flesh of a chicken used for food." The average American eats more than one hundred pounds of chicken per year, according to the National Chicken Council, making it the most popular meat consumed in the United States. Worldwide, some seventy billion birds yearly are roasted, boiled, Kentucky Fried, and turned into everything from McNuggets to the famous "Jewish penicillin," chicken soup. (One of my friends, an award-winning journalist and talented cook, was shocked to learn, when she was nearly forty, that it is physically possible to make soup without chicken stock.)

The disturbing fact that, on the way to the soup pot, a chicken can continue to run around after decapitation does little to bolster appreciation for the species' more refined traits. One rooster was able to live for eighteen months after his head was cut off. Farmer Lloyd Olsen, hoping to please his visiting mother-in-law, who particularly savored boiled chicken neck, failed to kill the rooster when his ax missed the bird's carotid artery and left one ear and most of the brain stem intact. Not only did the victim survive, but he grew from two and a half pounds to eight, and attained national fame as Mike the Headless Chicken on the sideshow circuit from 1945 to 1947. Even now, the rooster's hometown of Fruita, Colorado, holds the Mike the Headless Chicken Festival the third weekend in May each year—a day of races, games, and food intended, as its organizers say, to celebrate the bird's "admirable will to live." But alas, Mike's story also perpetuates the one "fact" most people claim to know about chickens: that they are automatons too stupid to know if they're dead or alive.

But, as I've come to learn over the decades of sharing my life with successive flocks of these affectionate, industrious, and resourceful birds, almost everything people "know" about chickens is wrong.

My friend Gretchen Morin began my education. When my husband and I bought the 150-year-old farmhouse we'd been renting in southwest New Hampshire, she gave us our first flock of hens as a barn-warming present. Gretchen raised organic vegetables and Connemara ponies on a farm near our house, and when she had moved there, years earlier, a friend had given her the gift of twelve hand-raised black hens. Gretchen had been delighted at the prospect of fresh free-range brown eggs every day, but the flock provided much more. "It was the most incredible gift anyone ever gave me," she said. "I had been given the gift of an entire world—a whole chicken universe."

Of course, back then, I had no idea what she was talking about. But I was eager for a flock of my own. Every animal I have ever known has bettered my life in some way. I love birds—I have lived with finches and parakeets, cockatiels and lovebirds, and a gorgeous red-and-blue Australian parrot called a crimson rosella—but I had never lived with a whole flock of them. What would the hens reveal? "You'll see," Gretchen promised. "Nothing has ever made me happier."

At the Agway feed store, Gretchen ordered for us twelve chicks of the same breed she'd first owned—Black Sex-Links, so named because the females can be identified upon hatching by their all-black color, averting the problem of raising a coop full of jealous roosters—and hand-raised them in a heated trailer on the farm. My husband and I would often visit them there, holding one or two peeping chicks in our hands, or on our laps, or tucked into our sweaters, speaking softly to each so she would know us. When they were old enough—no longer balls of fluff but sleek, slim black miniatures of their eventual adult selves—they moved into our barn. Our travels in the Chicken Universe had begun.

At first I was afraid they'd run away or become lost. We had a cozy, secure home for them prepared on the bottom floor of our barn, with wood shavings scattered over the dirt floor, a dispenser for fresh water, a trough for chick feed, some low perches made from dowels, and a hay-lined nest box made from an old rabbit hutch left over from one of the barn's previous denizens, in which they could, in the future, lay eggs. Chickens need to be safely closed in at night to protect them from predators, but by day we didn't want to confine them; we wanted to give them free run of the yard. But how could they possibly understand that they lived *here* now? Once we let them out, would they even recognize their space in the barn and go back in it? When I was in seventh grade, my family had moved, once again, to a new house; my first afternoon there I literally got lost in my own backyard. Could these six-week-old chicks be expected to know better?

Gretchen assured me there would be no problem. "Leave them in the pen for twenty-four hours," she told me. "Then you can let them out and they'll stick around. They'll go back in again when it starts to get dark."

"But how do they know?" I asked.

"They just do," she said. "Chickens just know these things."

When before dusk, I found them all perched calmly back in their coop, I saw that Gretchen was right.

In fact, chickens know many things, some from the moment they are born. Like all members of the order in which they are classified, the Galliformes, or game birds, just-hatched baby chickens are astonishingly mature and mobile, able to walk, peck, and run only hours after leaving the egg.

This developmental strategy is called precocial. Like its opposite, the altricial strategy (employed by creatures such as humans and songbirds, who are born naked and helpless), the precocial strategy was sculpted by eons of adaptation to food and predators. If your nest is on the ground, as most game birds' are, it's a good idea to get your babies out of there as quickly as possible before someone comes to eat them. So newborn game birds hatch covered in down, eyes open, and leave the nest within twenty-four hours. (An Australian game bird known as the mallee fowl begins its life by digging its way out of its nest of decaying vegetation and walks off into the bush—without ever even meeting either parent.)

That chickens hatch from the egg knowing how to walk, run, peck, and scratch has an odd consequence: many people take this as further evidence that they are stupid. But instinct is not stupidity. (After all, Einstein was born knowing how to suckle.) Nor does instinct preclude learning. Unlike my disoriented seventh-grade self (and I have not improved much since), young chickens have a great capacity for spatial learning. In scientific experiments, researchers have trained days-old chicks to find hidden food using both distant and nearby landmarks as cues. Italian researchers demonstrated that at the tender age of fifteen days, after just a week's training to find hidden food in the middle of their cage, chicks can correctly calculate the center of a given environment—even in the absence of distinctive landmarks. More astonishing, they can do it in spaces they have never seen before, whether the area be circular, square, or triangular. How? The chicks "probably relied on a visual estimate of these distances from their actual positions," wrote University of Padua researcher L. Tommasi and co-authors in the *Journal of Comparative Physiology*, "[but] it remains to be determined how the chicks actually measure distances in the task."

We never determined how our first chickens knew their new home was theirs, either. We never knew how they managed to discern the boundaries of our property. But they did. At first, they liked to stay near the coop. But as they grew,

Sy introduces Jane Cabot to the new chicks.

Revered for their soft, silky feathers and gentle nature, Buff Non-Bearded Silkies like this one make great pets and foster moms. Originally created in China, the breed has black skin—and black bones!

The Columbian Wyandotte debuted at the 1893 World's Columbian Exposition. The breed was created by crossing a White Wyandotte with a Barred Plymouth Rock.

"Croaky" here belongs to a group of chickens informally collectively known as "Easter Eggers"—because the hens lay naturally colorful eggs, which can range from pinkish to olive to aqua.

The beautiful Mille Fleur breed's name translates to
"thousand flowers" in French.

A Leghorn rooster crows. This breed originated in Tuscany and was first exported to North America in 1828.

When hens and roosters of different breeds choose their own mates,
a handsome mix like this proud chicken can result.

The Crested Appenzeller, a Swiss breed, looks fancy compared with the popular but much plainer White Plymouth Rock.

they took to following me everywhere, first cheeping like the tinkling of little bells, later clucking in animated adult discussion. If I was hanging out the laundry, they would check what was in the laundry basket. If I was weeding a flower bed, they would join me, raking the soil with their strong, scaly feet, then stepping backward to see what was revealed. (Whenever I worked with soil, I suspect they assumed I was digging for worms.) When my husband, Howard, and I would eat at the picnic table under the big silver maple, the Ladies would accompany us. When my father-in-law came to help my husband build a pen for Christopher Hogwood, then still a piglet, the Ladies milled underfoot to supervise every move. The hens were clearly interested in the project, pecking at the shiny nails, standing tall to better observe the use of tools, clucking a running commentary all the while. Before this experience, Howard's dad would have been the first to say that he didn't think chickens were that smart. But they changed his mind. After a few hours I noticed he began to address them. Picking up a hammer they were examining, he might say, directly and respectfully, "Pardon me, Ladies"—as if he were speaking to my mother-in-law and me when we got in the way.

But when their human friends are inside, and this is much of the time, the Ladies explore on their own. A chicken can move as fast as nine miles an hour, which can take you pretty far, and ours have always been free to go anywhere they like. But ours have intuited our property lines and confine their travels to its boundaries. They have never crossed the street. And for years, they never hopped across the low stone wall separating our land from that of our closest neighbor. That came later—and it was not the result of any physical change in the landscape, but the outcome of a change in social relationships among their human friends.

When the Ladies first moved in with us, Larry Thompson lived next door with his Airedale, Cooper, both of whom we liked and visited. When he moved out, the house sat vacant for a time. Still, the hens wouldn't venture over the low wall. Finally Lilla Cabot and her two blond, blue-eyed girls, Jane, seven, and Kate, ten, moved in. Understandably enchanted with our friendly black-and-white spotted pig, the girls visited the barn regularly, bringing treats (often their uneaten school lunch, which they had saved all day for this moment), petting him, and escorting him on daily rooting excursions. Next the girls were helping me gather the Ladies' eggs and tossing the Frisbee for Tess, the border collie. Soon we were together baking cookies, reading animal stories, and visiting back and forth daily. That's when the Ladies started hopping over the stone wall. Somehow they realized, before we humans did, that our two families had become one unit.

This should not have surprised me. To chickens, relationships are extremely important. Researchers have documented that an average chicken can recognize and remember more than one hundred other chickens. How? They may well remember a gestalt of features, including the voice. But facial features seem to be particularly important. Researchers A. M. Guhl and L. L. Ortman fastened fake combs on hens' heads, to discover that the new headgear rendered them strangers to their flock-mates. So did dyeing the feathers on the head. Dyeing the feathers new colors elsewhere on the body or even removing patches of feathers did not. Like us, birds seem to look into the faces of their friends. In his revelatory book *The Minds of Birds*, Alexander Skutch tells the story of a small bird of prey known as a kite, who had been fed by a particular soldier, Derek Goodwin, at an army camp in Egypt during World War II. Soaring above marching columns of identically dressed soldiers, the bird would find and hover above Goodwin and Goodwin alone when Goodwin looked up, revealing his face.

Because chickens live in flocks, the ability to identify individuals is even more important than it is to a bird like a kite. Belonging is essential to a chicken's well-being, as is clear from the complex social system of the pecking order. The pecking order is not always a straight hierarchy. In a study of captive red jungle fowl, chickens' immediate ancestors, a University of Florida researcher constructed "sociograms" diagramming the relationships between all the birds in each of four flocks. In many flocks, there were several groups of up to three hens who were not only friends but coequals.

The pecking order is more about order than pecking. Chickens do peck—sometimes to the death—but mine, over the course of many years and many flocks, have pecked with admirable discretion and restraint; sometimes an "air peck" or merely raising the hackles gets the message across. Such a gesture is delivered to remind another chicken of her position in the group. And dominance is not always established by bullying. Chickens seem to understand that with great power comes great responsibility. A dominant chicken serves as a peacekeeper among the flock, settling squabbles and looking out for danger.

It's important that everyone knows her place. When it comes to roosting at night, the pecking order determines who sleeps next to whom, and on which perch. It does not always determine who eats first, but usually predicts, if there is a squabble between individuals over a choice food morsel, who will ultimately win.

Within this well-established order, ours is a peaceful flock. But it is not immune to violence. A skunk (whom we later captured, moved, and released) dug in through the dirt floor on subsequent nights and killed and ate two hens. A fox carried off another. A wandering dog killed more. The little flock was shrinking. Though far longer than the mere five to eight weeks of life of the supermarket chicken, alas, the average natural life of a pet hen spans only five or six years—though Matilda, an ivory-colored Red Pyle bantam in Alabama, lived to the age of sixteen, earning her an entry in *Guinness World Records* and a spot on *The To-night Show.* But her impressive world record was usurped by Muffy, a Red Quill Muffed American Game hen from Maryland, who was twenty-three years and 152 days old when she died in 2011. By 2023, there was a new contender for the record. Twenty-one-year-old Peanut, a bantam hen who resides in Michigan, then claimed the title of oldest *living* chicken.

I thought sadly of the day my flock would be reduced to a handful of ancient, menopausal hens. But again, Gretchen knew what to do, and I have followed her advice: augment the flock by adding new babies I hand-raise myself.

They come in the mail—like a fruit-of-the-month order, or a book ordered over the internet. But this package is peeping when it arrives. I always tell everyone at the post office to watch out for my special delivery, and Mike or

Janet calls me the moment the order arrives from Cackle Hatchery in Lebanon, Missouri—a box of live baby chicks, hatched just two days before.

Lovingly I lift the perforated lid to a straw-lined cardboard box, not much larger than a big box of chocolates. The fluffy, peeping babies are still shaped like eggs. They'll never see the hen who laid the eggs from which they hatched. I'm their mother now, and I love them with a fierce tenderness that never abates.

In my home office where I write, I give the chicks free run of a big box that once shipped a refrigerator, carpeted with newspaper and wood shavings and warmed by a heat lamp. From the first time I did it, raising chicks in my office seemed perfectly normal to me; I showered in the morning with a cockatiel, slept with a dog, and spent many sunny summer hours lying in a field with a pig. Why shouldn't I have peeping chicks in my office?

Most of the day, at least one chick, often two, is somewhere on my body. For the next six weeks, until their baby down is replaced with feathers, I spend my days writing with a chick or two on my lap, beneath my sweater, on my shoulder or knee. Yanking at the tiny gold cross around my neck, or, worse, hopping onto the keyboard signals it is time to switch chicks. When I speak with strangers on the phone, they've been known to ask, "Are you calling from a zoo?"

Meanwhile, something magical has happened. Konrad Lorenz, the Nobel laureate credited with founding the modern study of animal behavior, called it imprinting. Students of animal behavior are careful to note that imprinting is not instinct; it is not learning; it is something of both. Most newly hatched game birds, including turkeys, ducks, chickens, and geese, will follow the first moving object they see, which is usually, of course, their mother. But working with hatchling graylag geese, Lorenz discovered that the babies don't instinctively recognize adult geese as members of their species. If a person is the first moving creature the gosling sees, the baby will follow the person as if he were the parent—as attested by many a charming photo of the white-bearded scientist walking down a path or rowing across a pond in his hometown of Altenberg, Austria, with a string of fluffy goslings following in single file behind.

Raising baby chicks in my office involves a great deal of cleanup. Unsightly blotches stain my clothing and dry in my hair. And when the downy chicks begin to grow feathers, every surface in my office—my books, photos, maps, notebooks, computer—is coated with a thick layer of powdery dust. (Each new feather grows in as a quill coated with a sheath of keratin, the same stuff as our fingernails. As the feather blossoms, the keratin breaks off in tiny pieces. Though this will happen again whenever a feather is replaced, never again is dust released in such quantities.) But it's well worth it. I adore these fearless, busy little souls, already so full of life and purpose. I am honored to follow in the footsteps of a great ethologist—as my chicks will follow in mine. And imprinting has a later benefit as well: human-imprinted babies later direct toward their person many of the inborn responses that normally would be shared only with a member of their own species. In this way, I become an honorary chicken.

Chickens with poufy topknots; chickens with feathered feet; chickens with turquoise earlobes; chickens who lay green eggs (though not with ham); chickens with tails that can grow twenty feet long . . . When I placed my initial order with the hatchery, I had many breeds to choose from. Chickens have been living with people for a very long time (by some reckoning, as long as eight thousand years— longer than donkeys and horses, longer than camels or ducks, and by some accounts, even longer than pigs and cattle). Starting with the red jungle fowl of Southeast Asia (who looks pretty much like the rooster on the Kellogg's Corn Flakes box), through selective breeding people have created as many as 350 different varieties of chickens— chickens spangled with iridescent feathers, chickens with naked necks (these are called turkens but are really chickens, not crosses with turkeys), chickens standing tall on long legs like basketball players, miniature chickens called bantams who might weigh only a pound.

Every winter we muse over the catalogs of the chicken hatcheries the way gardeners dream over seed catalogs. We glance at the bargains: one catalog carries a Top Hat Special, an assortment of crested breeds from the five-toed Mottled Houdans to the golden Buff Laced Polish, all of whom look like they are wearing giant, curly wigs made out of feathers. Even the babies sport little top hats. There's usually a special on Feather Footed Fancies: all these varieties have feathered feet, making the birds look a bit like they are wearing floppy slippers. There is even a Fly Tyer's Special, a selection of chickens whose feathers can be used to fashion particularly attractive fishing lures. Of course, we ignore the Frying Pan and Barbecue specials; I'm a vegetarian and certainly wouldn't eat anybody I know.

What we're really looking for are handsome, vigorous chickens who do well in cold climes. With their glossy black feathers, red, upright combs, and ample bodies, our Black Sex-Links were all these things. We sometimes called them "the Nuns," especially when they raced out of their coop each morning like a flock of chatty Sisters leaving a convent in their billowy black habits. Such a uniform appearance did they present that, without noting the subtle differences in the shape of their combs, it was nearly impossible to tell them apart. (Scientists faced with this problem sometimes outfit the hens with numbered armbands affixed to a wing.)

Adding birds of different breeds presaged an important change in our understanding: now that it was easier to tell birds apart, the distinct personalities of individuals began to reveal themselves more clearly.

Kate and Jane next door took this opportunity to give the chickens names. I had not done so before because of a military adage learned from my father, an army general, which warned, "Never name the chickens." I knew this wasn't about poultry, but about the commander's responsibility to remain objective about the troops he must choose to send into battle. But somehow it always stuck with me that something bad would happen if you named your hens. In fact, the only one of the Nuns who'd been named before the arrival of the new babies bore the badge of near catastrophe: the girls named her Foxy Lady after an encounter with a fox left her with no tail feathers. But with the coming of age of our creamy buff Speckled Sussexes and our black-and-white Lakenvelders came more names: Snow White for one of the "Lakes," who was exceptionally beautiful and loved to fly; Madonna for a loud and theatrical Speckled Sussex; Matilda for one of the remaining Black

Sex-Links, who, now older, walked with a rolling gait resembling a waltz. . . .

Soon it became evident that some hens were consistently outgoing and others shy; some were loud and others quiet; some cautious and others reckless. This was particularly obvious whenever the hens faced a threat, such as a hawk flying overhead. Some hens hid in pricker bushes; others raced inside the coop. Some dashed behind a large board that leaned against an outside wall of the barn. Some individuals would continue to hide for more than an hour. They were so good at hiding that sometimes, alerted by their alarm calls, I'd rush outside to try to defend them and spend half an hour trying to find them and entice them from their hiding places. Brassy Madonna was often first to emerge, while Foxy Lady, surely remembering the horrible fox, stayed immobile the longest. But all of them understood that even though the hawk might not be visible, it still might be lurking somewhere nearby.

Chickens both remember the past and anticipate the future. This has been clearly demonstrated in the laboratory. In one study, published in the journal *Animal Behaviour*, Silsoe Research Institute biologist Siobhan Abeyesinghe and her co-authors tested hens with colored buttons. When hens pecked at a particular button they were rewarded with food. But they got an even bigger reward if they learned to postpone their pecking. If they waited—for up to twenty-two seconds—they got even more to eat. The birds chose to wait for the jackpot more than 90 percent of the time.

Experiments like this show that chickens "can do things that people didn't think they could do," says Christine Nicol, professor of animal welfare at the University of Bristol in England. "There are hidden depths to chickens, definitely."

In our attempt to plumb those depths, the girls and I tried to decipher the chickens' language. At first my husband dismissed our efforts, insisting that most of what they were saying was, "I'm a chicken. You're a chicken. I'm a chicken." He gave them more credit than most scientists did for many years. Even though birds have the greatest sound-producing capabilities of any vertebrate—far superior in both volume and range to the greatest human opera star—their distinctive calls and elaborate songs were not considered true communication. Even parrots who spoke sensible phrases in English to their human owners were dismissed as mere mimics. Birds' spectacular voices were merely unconscious, uncontrolled noises reflecting the birds' inner states (which were also assumed to be unconscious).

The ancients knew better. The word "augury" comes from the Greek word meaning "bird talk," for to understand the language of birds was to understand the gods.

And the Cabot girls and I knew better, too. We could feel the anguish in the Ladies' calls when they spotted a predator; we could read their delight when someone found a mother lode of worms or beetles in the compost pile. We discovered, too, that some hens announce the blessed moment when they have laid an egg: a loud, measured series of rising *buk-buk-buk-AHH!* sounds. We suspected this meant more than just "Ouch!" Hens may perceive their eggs as gifts that may be presented to their rooster, their flockmates, or an honorary chicken/person. On Farm Life Forum's Web page for poultry keepers, a woman wrote of a chicken her father had kept as a pet when he was a boy. Each evening, the hen appeared at the door of the house and would peck to be let inside. When the door opened, she would proceed directly to the boy's bed—where she would lay an egg on the pillow. Then, gift delivered, she would stride back to the door and return to the henhouse.

New research published in the journal *Royal Society Open Science* shows that ordinary humans—even nonscientists with little or no previous experience with poultry—can intuit a chicken's emotional state by listening to their clucks. Nearly 70 percent of people could tell the difference between the voice of a happy, excited chicken and that of an unhappy one just by sound alone, without even seeing the bird.

But chickens' voices express far more than pleasure or discontent, of course. At Macquarie University in Sydney, Australia, working with Golden Sebright chickens, a breed whose voices are most similar to the ancestral jungle fowl, psychology professor Chris Evans and his wife, Linda, have identified twenty-four distinct calls the birds use to communicate specific information to others in the flock. For instance, playbacks of a rooster's kissing *took-took-took* call caused hens to search for food—evidence it means "Come, here's some food" and not merely "I'm happy." Critics countered that hens hearing the call were only succumbing to a knee-jerk reaction: the call was a trigger causing hens to peck the ground mindlessly. To test this, the researchers divided the hens into two groups: one got a snack just before hearing the food call. The others had none. The hens responded as humans would to language. Those who had just eaten showed limited interest, but those who were hungry searched the ground for food. "If you're on a long drive and you pass a restaurant sign, that could be a salient piece of information. But if, after food has been brought to the table, someone says, 'There's food,' that's a redundant comment. It's that kind of contrast," Chris Evans explained.

Not only do hens understand when a call is about food; they can even discern from a rooster's call what's on the menu. The researchers reported that the rooster called at a faster rate if the food discovered was especially tasty—like their favorite, corn, instead of the layer mash ration that is their regular feed.

The Evanses also found that chickens used several different alarm calls, depending on the size, shape, speed, and location of the predator. The researchers mounted a video monitor in the chickens' cage on which they could project the images of various predators in different conditions. A video of a hawk prompted a high-pitched scream, delivered while the bird crouched. A video of a raccoon elicited a pulsating series of ten high-pitched *buk* sounds, followed by an alarmed *AH!* while the bird paced about. When these calls were recorded and played to other chickens, the others clearly understood what they meant: the high-pitched scream made them scan the sky, while the agitated clucking prompted a search of the ground. The alarm calls were more vehement when the predator was nearby or approaching quickly—and in the case of the hawk, delivered more frequently when the rooster knew he had an audience. (Apparently, a ground-predator call is not only intended to warn hens but is addressed to the predator as well—probably to let the predator know he's been seen.)

In 2017 I was privileged to meet the author of a book actually called *How to Speak Chicken.* Melissa Caughey, who lives with her flock in Cape Cod, and I met at an American

Association for the Advancement of Science event in Washington, DC, where she was getting an award for her book. In one of our conversations, she mentioned that several of her hens utter a very distinctive sound when they see Melissa coming toward the coop. She had a recording of it: three low sounds (*bup, bup, bup*) followed by another one (*BAAAH!*) an octave higher. She had been hearing some of her hens do this since 2011, she told me, and thought it was a greeting. To me, it sounded like an announcement—rather like a trumpet's blast before the arrival of a queen. Melissa realized that her hens had given her a name.

A number of other animals assign names to each other. It's useful. How else to call, find, and refer to one another? Dolphins have distinctive signature whistles for each member of the pod. Venezuelan green-rumped parrotlet parents give each chick in their brood an individual name. And each member of a clutch of Australian fairy wrens must learn their family's unique surname. The name, which both parents use as well, functions like a password. If the nestlings don't say it, they won't get fed. (The strategy helps parents avoid feeding nest parasites like cuckoos, who lay eggs in other birds' nests.)

Once Melissa and I returned from our travels, we delved deeper into her discovery. She had recordings of her flock dating back to 2011. Melissa determined that her most vocal and lead hen—Tilly, in whose honor her excellent blog, *Tilly's Nest*, is named—coined the name, and the several others picked it up from her. "Now that she is gone," Melissa wrote in her blog, "my name is even more special.

"I think sometimes, we spend our lives doing too much of the talking," she concluded her blog post about her chicken name. "Animals have a lot to say."

They have a great deal of wisdom to share with us, as well as comfort and joy. The hen conversations I love most are spoken in sleepy voices, as the Ladies get ready to roost for the night. I check on them each evening and turn out their light. In summer, when the sun sets late, hens are often still milling around the floor. I cry, "Perching!" and in response they fly to their regular roosting spots, each bird surrounded by her closest friends. They settle in for the night, making their long, low, contented nighttime chatter—agreeing, no doubt, that in the Chicken Universe, all is right with the world.

It is easy to believe them. When hens are calm, nothing is more soothing than their voices, especially when punctuated by the occasional grunt of a sleepy pig. Sometimes, lulled by their cozy, restful sounds, I lose track of time, enveloped by a sense of belonging, washed in peace and wholeness among a sisterhood of hens. Some evenings my husband finds me in the henhouse, caressing one or two

chickens, eye level with my perching friends, as if one of
the flock. He has sometimes overheard me join their evening conversation. "Yes, Ladies," he heard me say one
night, "you're my beauties. I love you so much."

Some of the most memorable of the many quirky chickens
we've known have been roosters. Only once did we actually order them: seduced by the promise of their glorious
long tails, we ordered and paid for two lovely Lakenvelder
cockerels.

Our other roosters, though, arrived unbidden. In my
earlier days of chicken husbandry, I ordered from a hatchery
that, as a bonus, included in every order a "free exotic chick."
It might be an Araucana, the kind that lays green eggs, or a
stately Blue Andalusian, with its upright posture and blue
feathers. It might be an Egyptian Fayoumi from the Nile, or
perhaps a Cuckoo Maran, a French breed laying chocolate-
colored eggs. . . . It was always exciting to see who turned up
in the order. We'd have to wait until the feathers came in to
identify the breed with which we had been blessed.

Invariably, though, it would turn out to be a rooster.
Always he turned out to be quite handsome; and usually,
one day, he would turn on us.

We had heard about this problem. A neighbor's boy had a

wonderful rooster as a pet. He used to ride on the handlebars of the kid's bicycle. But then one day, he turned. He attacked his former buddy relentlessly, flying at the boy's face with his sharp spurs. The little boy was bruised and bloodied every day. The parents kept the child, but gave the rooster away.

We felt sorry for the family, but we considered the incident a fluke. Surely nothing like this could happen to us—not with all our chicks firmly imprinted.

Besides, we were thrilled to have roosters. We didn't need a rooster to get our hens to lay (though only fertilized eggs will hatch), but a rooster has much to offer. Hens can hope for no better protector. We can't be in the yard with them every minute, but a rooster can, and he will fight to the death to protect his flock. Flinging himself spurs first at his opponent, a cock will fight with such ferocity and determination that the ancient Greeks believed even a lion would fear him.

Most roosters are very solicitous of their hens. When he's not patrolling for predators, he's often searching for food his flock might enjoy. After he finds it, he utters the food call that the Evanses studied, then stands aside while his women enjoy the treat, and only after they've had their fill will he sample the snack. The Talmud praises the rooster, and its writers advise Jews to learn from him courtesy toward their mates.

We eagerly awaited our cockerels' transition to maturity. When our roosters began to crow, we loved it—once we figured out what was happening. One day I heard an unearthly racket, a sort of strangled gargling, coming from the backyard and rushed outside, fearing I'd find a child or some small animal who had been injured. There was our rooster, Clarence, on the stone wall, standing tall and proudly practicing his crow. Like an adolescent singing, it takes a rooster some time to find his mature voice.

Luckily, our neighbors didn't mind the crowing. The Cabot girls and their mom were delighted, and almost everyone else on the street had kept chickens of their own at one time or another. The crow of a cock is a part of the soundtrack of rural life. In the sacred book the Hadith, the prophet Muhammad tells us why roosters crow: they do so because they have seen an angel. The moment a cock crows, the holy man advises, is a good time to ask for God's blessing.

And so is the first time your rooster goes on the attack.

We were utterly unprepared. Our first roosters, the two Lakenvelders, had been such gentlemen, keeping largely to themselves, tending their small flock of seven Lakenvelder hens, mostly ignoring the others. But when our first "free exotic chick" turned into a rooster, that was another story.

Alex the Araucana had been a bold, cheerful chick. A

born leader, he would help me round up the other chicks if I needed to close up the coop before dark. Because he was bigger and bossier than the others, we suspected from the start that he might be a rooster and thought he would make a fine one. When he started to crow, we knew. But we didn't know yet what darkness lurked in his rooster heart.

One day, Howard was lying on the ground, trying once again to fix our ancient secondhand lawn mower. From the inches between the machine's steel belly and the soft green grass below, Howard caught sight of something moving fast—malevolently fast—directly at him. Realizing that lying prone on the ground was not a good way to face an attacker, Howard leapt to his feet. Alex pulled up short right in front of him, as if he had come to his senses. "But I knew," said Howard, "that he was up to no good."

We hoped this was just a phase. That it was not was demonstrated by no less an authority than the minister at the Congregational church where I was a deaconess.

It was a rather momentous visit. Not only was Graham Ward my minister; we were friends, and I had been especially close with his wife, who had died tragically of cancer. But eventually, Graham found a new lady love. One day he brought his bride-to-be, Kathy, with two of her small children, over to meet our charming pig, our Frisbee-playing border collie, and our affectionate flock of chickens. It was meant to delight both the lady and her children, rather like a visit to a petting zoo. But it didn't turn out that way.

The pig, then about five hundred pounds, bucked and thundered out of his pen, full of porcine exuberance. Kathy was visibly alarmed. "I didn't know he would be that big!" she exclaimed. Christopher ran to his main outdoor feeding area, and we hooked him up to a tether so the kids could watch him eat a bucket of slops. (When it came to eating, he was a performance artist.) His large tusks did not escape the young mother's notice. She took her daughter's hand. Though her son was fascinated, I was afraid the little girl might cry.

Now, as if to our rescue, the Ladies rushed over to share in the food bounty. "Here's someone more your size!" Graham said to the little girl. "Watch—you can pat her." As he had done many times with our hens, Graham, dressed in shorts, reached down to pet one of them.

Just like when I stroke the Ladies, Graham started to run his hand down her back. And just as the hens always do with me, she assumed her distinctive squatting posture. This is a well-known chicken behavior usually directed at a member of her own species. It is actually known as a "sex crouch." It's a position that a chicken normally uses to make it easy for a rooster to mount her.

Just then, from another part of the yard, Alex the rooster looked up. The scene looked innocent enough to human eyes:

two small kids, three adults, a pig, several chickens, all standing around together on a sunny summer day. But through his rooster eyes, Alex saw a moral travesty, an insult to his roosterhood: the minister was trying to have sex with his hen.

I only had time to cry "Graham, watch ou—" before Alex hit. Spurs first, the enraged rooster flew at the back of Graham's bare calves with the full force of his fury. He broke the skin. His fiancée was of course aghast; the little girl began to cry. Graham tried to assure the children the rooster wasn't mean; he was just protecting his flock. But the kids were having none of it. This was "Nature, red in tooth and claw." I carried Alex ignominiously away to the coop, while Graham, bloodied, ushered his new family to the car. I tried to end things on a good note by handing the shocked children each a fresh egg to take home. Unfortunately, I later learned that one broke on the way back, staining the car's upholstery.

None of our roosters stayed for long. The gentlemanly, long-tailed Lakenvelders, always pictures of vigorous health, dropped dead from their perches within days of each other before their second birthday. Gretchen told us this is not uncommon. Roosters' constitutions are quite different from those of hens, right down to a marked difference in respiration: a hen breathes thirty to thirty-five times a minute; a rooster only eighteen to twenty times. Apparently their different physiologies make the males more likely to drop dead, a phenomenon we named Sudden Rooster Death Syndrome.

The other roosters, we had to ship off. A farmer a few towns over was glad to take them and assured us they weren't destined for the pot. "He's a good 'un!" he said, holding one of them upside down by his feet. The farmer raised exotic chickens and needed roosters for his many hens.

The Ladies, frankly, seemed somewhat relieved. Though they had appreciated their roosters' food calls alerting them to particularly juicy worms or hidden treasure in the compost pile, there was a cost: the regular annoyance of someone jumping on your clean back with his dirty, scaly feet and biting your comb with his beak— whether you felt like it or not. Sometimes our hens would squawk in annoyance.

Free of roosters, the flock seemed more composed, cohesive, and affectionate. The peace of the henhouse was restored—a feminist utopia.

But even our gentle Ladies have a strange and sometimes disturbing side to their souls. A Speckled Sussex whom the girls named Pickles—the only hen who enjoyed this particular food—revealed to me that the Chicken Universe, though in many ways the sweet soul of domesticity, is inhabited by aliens.

Pickles was a special needs chicken. The moment she arrived as a fuzzy chick we saw she had a bump at the top of her head the size of a small pimple. We soon discerned, with growing dismay, that she had a tiny hole in her skull and that the pimple was filled with cerebrospinal fluid. But she seemed otherwise healthy and normal.

As Pickles shed her baby down for feathers, we began to realize she was a little slow. She seemed to be the last one to notice a food call, for instance, and when the other chickens would come running, she usually brought up the rear. She didn't react normally to loud noises or quick motions: unwisely, she accompanied Howard when he was sawing wood, her neck, I thought, perilously near the blade. Pickles had a tendency to wander off on her own, though not in the spirit of adventure. She always seemed a little lost.

At first, we didn't realize that Pickles couldn't peck straight. When we put down feed, there was usually enough of it that if a hen pecked anywhere within a half-foot radius, she'd hit something to eat. Only later, when we would offer Pickles a small, single morsel from our hand—a tiny ball of pie dough, a piece of carrot—did we perceive her problem. She'd consistently peck about two inches to the right of it, entirely missing the treat.

But no one in the flock ever minded that Pickles was a challenged chicken. We suspected she was the lowest in the pecking order, but nobody had to peck her to drive the point home. Pickles accepted her station without question; she had no ambition to rise in the hierarchy, so there was no reason for quarrel. If anything, Pickles's problems made Kate love her all the more. She'd carry the brown hen around with her everywhere, sometimes tucking her beneath her sweatshirt or sweater, and Pickles would remain perfectly calm and contented there.

One day Howard and I heard a chicken commotion in the yard and rushed out to see. Somehow, on her wanderings, Pickles had sustained a hideous injury. A long slash stretching perhaps four inches along the length of her neck—had she run into barbed wire?—had nearly severed her head. But what horrified us even more was the way the rest of the flock greeted their wounded comrade. Our Ladies, normally such paragons of calm and homey affection, were biting cruelly at the bloody wound, scolding and chasing her savagely.

What was happening? Why would the Ladies do such a thing?

"It's just something chickens do," Gretchen later told me, almost apologetically. "It seems so mean. But they can't help it." That chickens will attack and even kill a wounded flock-mate is so well known that products have been developed to cope with it. These have included chicken eyeglasses—the first models were invented in 1903, and some were still in use as recently as the late 1970s. Many of the glasses were rose-tinted, which prevented the birds from seeing red. Another product available at most ag stores is called Stop Peck. It comes in a bottle like Elmer's glue, looks like clotting blood, and to chickens, apparently, tastes awful.

But I didn't know that then. All I knew was Pickles needed help fast. I scooped her up and held her in a box on my lap as Howard drove us to the vet.

Our wonderful vet, Chuck DeVinne, cleaned and disinfected the wound and gave me some antibiotics to mix in her water. She'd have to be separated from the flock, he told me. Until the wound healed, it would be a target impossible to resist. And in fact, the most immediate threat to her recovery was this same deep, ancient chicken instinct: to peck at the sight of blood is a drive so strong that a chicken will often kill herself doing so to her own wound.

Chuck fashioned her an Elizabethan collar from a discarded X-ray, wished us luck, and sent us home.

"There is a deep vein of locked-in behavior in all birds," says Gary Galbreath, an evolutionary biologist at Chicago's Field Museum and a professor at Northwestern University.

Birds are thinking, feeling creatures, but some of what they do is beyond their conscious control, irresistibly carved into their genes.

Lorenz's fellow Nobel laureate, Niko Tinbergen, showed this with a series of striking experiments with herring gull chicks. Normally wild chicks peck at the red spot on the parent's bill to induce the parent to regurgitate food. But they will also, he found, peck at a red dot on a fake "bill" attached to a disembodied head made of cardboard. Tinbergen and his students presented the chicks with a series of options. They found chicks will peck at the red dot even if the head is painted blue. They'll peck at the dot even if the head is weirdly misshapen. And they will preferentially peck at a giant red dot instead of the bill of a real parent if given a chance—even when pecking the giant red dot never produces any food for them at all. For the chick, the red dot acts almost like a push button does for a machine, "releasing" an inborn, preprogrammed pecking response.

In another experiment, Tinbergen showed that a mother oystercatcher, a shorebird, may make an equally strange choice. Normally she lays a clutch of three eggs, but she may lay up to five. But once she begins to lay, if presented with a giant fake egg—painted in natural colors—she will abandon her own. She'll rush to the giant egg, frantically trying to sit on it, every time. She cannot resist.

Konrad Lorenz even observed such mindless behavior in his tame jackdaws, crow-like birds who sometimes demonstrated their affection for him by bringing worms to his mouth (when he refused to eat them, the birds would resolutely stuff them in his ears). But if he took any one member of his flock into his hand, the others would attack him. One day, quite by accident, Lorenz made a profound discovery. After coming back from a swim, he was standing on the roof among his jackdaws when he remembered his black bathing suit was in his pocket. The minute he took out the limp black clothing, his jackdaws suddenly flew into a panic, shrieking alarm calls and attacking him with their feet and bills.

It was not that these birds mistook the swimsuit for a fellow jackdaw. Birds' eyesight is excellent, and they were at close range. There was no misunderstanding. There was no understanding at all. The sight of something limp and black triggered the behavior. The response was as involuntary as a knee jerk.

"Such 'mistakes' . . . have always been emphasized," wrote Tinbergen in his landmark *The Herring Gull's World*, "and it was usually pointed out that 'instinct' was 'blind' and rigid, and made the animal behave very stupidly when confronted with unnatural situations." But this interpretation, he warned, is "inadequate and incorrect." Herring gulls are excellent learners; the intelligence of jackdaws is well documented. Yet sometimes, smart birds, capable of reasoning and forethought, are governed by an ancient, genetically determined program beyond their conscious control. Bird behavior is the product of both.

Sometimes birds' actions might look stupid or cruel to us. Even Konrad Lorenz was appalled when, hoping to breed a cross between a turtledove and a ring-necked dove, he put these two symbols of peace together in a spacious cage, only to find two days later that the female had flayed the male within an inch of his life. (Of course, had they not been caged, the victim could have escaped relatively unscathed.)

But we err to judge on the basis of an instance or two; the bird lineage has made its choice over millions of years. Over that vast span of unforgiving time, what might seem in an experiment like a silly mistake has proved to be, instead, a course of action so right, it means the difference between life and death. At times, it is better to let one's ancestors make the decisions—and the ancestors have left their instructions in the genes. It's a deep wisdom of a sort humans seldom recognize.

Why should chickens attack at the sight of blood? I haven't found a fully satisfactory answer. Possibly because the sight of blood usually signals meat, a rare treat, which should be quickly eaten. Perhaps because a severely wounded flock-mate might attract predators—better to drive away one member than endanger the whole group. I don't know. But although dismayed by my Ladies' inexplicable savagery, I was not angry. I was awed—reminded how privileged I am to be allowed to travel in this otherworldly, avian universe.

Pickles made a quick recovery, exiled to the downstairs bathroom, where she perched on the rim of the sink—a situation that alarmed dinner guests who did not expect to find a chicken in an Elizabethan X-ray collar watching them when they used the toilet. Fully healed, she was accepted back into the flock without a problem. Life with the Ladies resumed its peaceful, cheerful course.

The flock gained new members as we tried new breeds: Dominiques, the black-and-white ancestors of the Barred Rocks, and Black Australorps, a cold-tolerant Australian breed quite similar to our original sex-links. We lost some old birds to age and to visiting predators: a mink, an ermine, a hawk.

Equally sad for us, the little girls and their mother moved away, to be nearer their grandmother in Connecticut. Though the house next door sat empty, our hens continued to consider that yard theirs—but not the yard across the street. Crossing the street may never have entered the minds of the new chicks; if it did, perhaps the older birds dissuaded them. The boundaries of their territory seemed to be part of the flock's cultural memory.

New neighbors moved into the empty house. At first we were nervous—what if they had dogs or kids who would chase our Ladies or harass our pig? What if they didn't like our compost pile, which was convenient to our barn but only yards from the boundary of their property?

But we needn't have worried, for they turned out to be Bobbie and Jarvis.

In their previous home in upstate New York, the couple had raised three sons, numerous pigs, and several flocks of chickens. Bobbie told me her chickens' individuality and intelligence had taken her by surprise. In the summer, the hens liked to hang out beneath the kitchen window, and Bobbie was amazed to discover why. She used to keep a classical station on the radio when she worked in the kitchen, which they could hear through the open window. Apparently, they loved music. When the radio was not on, they preferred to forage near the picnic table. All except Leticia, a pretty White Rock, whom Bobbie had named after her great-great-great-grandmother. Leticia held herself aloof from her flock-mates, choosing instead

The gorgeous Silver Sebright breed is named for its early 1800s originator, Sir John Sebright.

The Wheaten Old English Game bantam originated in Great Britain.
Bantams as a group refers to very small chickens, and you can keep more
of them comfortably in a smaller space.

This young female (or pullet) Silver Penciled Cochin bantam
enjoys the heavy feathering of her breed to keep her warm in
New Hampshire winters.

One of the oldest and most popular chicken breeds,
the Speckled Sussex.

A White Bearded Silkie sports the poufy headdress that lands this breed in the "Top Hat" section of the chicken catalog.

A New Hampshire Red steps out proudly. The breed was developed
from selectively breeding Rhode Island Reds.

The White Polish breed is a showstopper with its full white feathering and impressive head feathers.

Ashley demonstrates how to calm a rowdy rooster:
just cradle him in your arms.

to forage alone—unless she could be with her human family.

The Coffins' hens had been tended by a series of handsome and courageous roosters, whose photos they still keep. The last of them had given his life for his flock, trying, unsuccessfully in the end, to defend them from a marauding dog. Shortly after the tragedy, Jarvis retired from his job brokering waste fibers to paper companies, and the couple moved here.

Immediately, Bobbie and Jarvis volunteered to help look after our pig and chickens. Beautiful, slender Bobbie loved to bring the hens treats. Soon she was buying cracked corn at the Agway just to feed our hens. The couple visited our pig and chickens so often that Jarvis, handy and enormously strong, built a wooden walkway over the boggy area between their yard and ours.

One day, we returned from a three-day trip to discover that Jarvis had replaced the rabbit hutch we'd used as a communal nest box with individual nest boxes covered with a long, sloped, hinged lid. Another time he rebuilt their perches, offering varied shapes and thicknesses for greater choice and comfort.

None of this kindness was lost on the Ladies, of course. The area immediately surrounding the barnyard provided the hens with an excellent vantage point from which to observe their benefactors. (Our barn is in fact closer to the Coffins' house than to ours.) The Ladies soon figured out where Bobbie and Jarvis kept the cracked corn. They listened for the distinctive slam of the Coffins' porch door. They watched for any sign that either of them—or any of their visiting grandchildren—might be coming over. If the Ladies were occupied elsewhere and failed to notice, Bobbie would call them, and they'd come running over. Jarvis would call individuals by name. After the Cabot girls left, he had taken over the chicken-naming responsibility: he named one older Dominique hen Mother Hubbard because she liked to look after new chicks and named an Australorp he considered particularly bossy Hillary, because Jarvis was an ardent Republican.

From their second-story back porch, the Coffins had an excellent view of our barnyard. But when we were away and the hens were under their sole watch, that wasn't good enough for Bobbie. To guard against predator attack, Bobbie used the same device she used to surveil sleeping grandbabies when they came to visit: the baby monitor. Any commotion would alert her to trouble and she could come running. Howard and I thought this was a great idea, and we bought one too. Not only did it ensure the flock's safety; as I wrote my books, my words and thoughts would be bathed in the soothing sounds of the calm chicken voices in our barnyard.

But those days would soon give way to a new era. Our lives would change radically with the arrival of more new neighbors: the Chicken Whisperer and her Rangers.

Elizabeth Kenney came recommended by a close friend and fellow animal lover. We needed a tenant for the half of our house we rented out as an apartment; Elizabeth was getting a divorce and needed a new place. But not just any place would do: it had to have a barn.

Elizabeth, thirty-six, worked with our friend as a record keeper for a local home care and hospice organization for the elderly. When I met her, I could see she must be very good at this: she's smart and empathetic, with an air of calm that instantly puts you at ease. But her job, as much as she enjoys it, is not her passion. Her passion is her chickens.

After we first met, Elizabeth sent me their photos. The flock of twelve included Barred Rocks, Rhode Island Reds, and Black and Red Sex-Links. In the divorce, she got custody of the flock, which she called the Rangers.

There was plenty of room for them in the barn. Two summers before, our pig, fourteen years old, had died of old age in his sleep; Elizabeth could move the Rangers into his old stall. Her dedication to the flock was evident the day she arrived that October. A friend came to help her erect a prefab chicken house within Chris's old pen. The builder of the coop had once been part of a carpentry team featured on the TV show *This Old House*, and the structure featured pane-glass windows, front steps, a linoleum floor, and a ramp leading to two tiers of nest boxes, curtained for maximum laying privacy. The perches were covered with soft towels that Elizabeth laundered every week. The walls of the coop were bedecked with artwork—pictures of chickens, of course.

Elizabeth's side of the house was soon decorated in similar fashion. The upstairs windows were curtained with fabric printed with hens. Visitors were welcomed with a chicken doormat. Chicken photos adorned the walls, including one of a chicken dinner—hens feeding happily from porcelain plates atop a candlelit, tableclothed dinner table. It appeared on a recent Christmas card, for which she had won an award. When I came over for my first long visit, she pulled out a thick album, labeled *The Gang: May 05*, so she could show me her first flock's baby pictures. (More chicks came the next year, and they were similarly documented, to make up her current flock of twelve.)

"Here's their first dust bath!" Elizabeth said. Like many birds, chickens love to roll, kick, and fluff their feathers in powdery dust. They enjoy this so much that when they're about to take a dust bath, they emit the same call—*kuk-kuk-kuk*—that they do when anticipating eating a treat. Turning a page: "Here's Peanut, that's Averil, there's Jan. . . . I just love Jan's face. . . ."

Every one of Elizabeth's chickens is named. To her, the identity of each is as obvious as a person's. For instance, "My Jan is nothing like Goldie," she told me. "Goldie's sweet as pie, hyper and curious, and maybe not too bright. Jan's so smart. Jan will stand and sometimes she'll stick her head in my pocket to get the bread out. She knows where it is. She's a wise old lady—an old lady in a young girl's body. She's number one in the pecking order, but she's not aggressive up there. It's like she's got her silent power and only asserts it when somebody really gets out of line, and usually it's just Nikki. And Nikki's second in line. . . ."

Elizabeth spends at least an hour a day playing with her hens on weekdays, and spends much of the weekend with them. She knows each chicken's distinctive voice. "I can tell, if I'm turned away from them, who's saying what," she told me matter-of-factly, "and a lot of times what they're doing when they're saying it. If they are making curiosity sounds, I'll know whose voice it is, and if she's happy or not. I can tell you Peanut is going to walk over toward the corncob, and Nikki's going to give her a nasty growl, and then Peanut's going to run that way. . . ."

Shortly, I would see that this was true, and more. Some of her chickens, I would discover, say certain things in a special voice, and use it only when speaking with Elizabeth. Others perform tricks on her command. I began to realize with growing amazement whom fate had brought to us as a tenant: we began to call Elizabeth "the Chicken Whisperer."

It's a warm, overcast autumn day, the close of an Indian summer. Elizabeth has just gotten home from work, and the Rangers are out. Most of the day they spend outside, behind the turkey-wire fence she has erected by their coop. In the afternoon, they listen for Elizabeth's car, whose sound they recognize, and know she'll let them out. Now, says Elizabeth, "they're ready to rumble!"

Elizabeth Kenney and the Rangers

Nathan Townsend

Elizabeth issues her "Come to me" call—it sounds like a slower version of the tufted titmouse's whistled *chiva-chiva-chiva*—and they come running, wings out and beating to propel themselves forward quickly.

"C'mon—charge! Go get 'em!" Elizabeth, as usual, has brought treats. She sits down on the lawn and holds out an apple. Wide-eyed Goldie, second lowest in the pecking order, arrives first, hops on Elizabeth's black pants, and starts stabbing the apple with her beak. Next is Rudy, the Red Sex-Link who often escapes from the pen during the day, and after her comes top-ranking, smart Jan, a Rhode Island Red. Jan draws herself up tall above Rudy, then looks down her beak and pecks her meaningfully before advancing upon the apple. Now Lola, a Black Sex-Link, comes running, and next Peanut, a Barred Rock.

Peanut is possibly Elizabeth's favorite chicken, if she can be said to have a favorite. When Peanut was a year and a half old, she developed a blockage in her crop—the muscular compartment where birds store and soften their food before it moves on to the gizzard, the main organ of digestion. When medicines failed, Elizabeth begged her vet to operate on the chicken. The first surgery failed. The second was a success.

During her long recuperation, Peanut lived in a spare room in the house, and Elizabeth fed her twice a day in the kitchen with a syringe. "Every morning, when I'd get breakfast, she'd hear me and purr like a cat," Elizabeth told me. "She'd fluff up and greet me and be very happy." Every night, as Elizabeth stroked her, Peanut would sit in her lap. When she bent over the hen, wisps of Elizabeth's shoulder-length auburn hair would dangle down. Peanut would take the errant strands in her beak and gently tuck them behind Elizabeth's ear.

The special closeness the two shared as Peanut recuperated still persists. Elizabeth bends down to kiss Peanut on the head. In response Peanut smacks her beak, making a kissing sound. "She only does this with me, only when I kiss her," Elizabeth explains. Lacking lips, Peanut is doing her best to kiss back.

Several other hens say things in a special voice, only to Elizabeth, never to other hens. "C'mon, Janny, c'mon, boit-boit!" she calls. Jan hops onto Elizabeth's outstretched legs. "Boit-boit-boit!" Elizabeth whispers into her ear. Jan cocks her head as if thinking for a moment. "Boit-boit-boit," the hen replies softly.

Another hen, Elizabeth tells me, says "duff-duff-duff." She'll show me later—but first she wants to demonstrate Jan's favorite game, which the hen herself made up. "Watch this," Elizabeth says to me, and then turns to Jan: "Too-too-too-too-too-too!" Jan seems to know what this means; she looks at Elizabeth expectantly, first with one eye, then the other. Elizabeth drops a small piece of bread from her pocket in front of the red hen. Jan picks it up in her beak. But she doesn't eat it. Instead, she ostentatiously drops it, then raises her head to broadcast loud clucks. "She's calling the others, just like a rooster would," Elizabeth explains. Hens look up from all around and start running. Goldie, still close by, is the first to arrive. She spots the treat—but before she can reach for it, Jan pecks it up and swallows it. That's the game. "She thinks this is really funny!" says Elizabeth.

The hens have other games as well. Elizabeth has brought a newspaper to demonstrate. "Whenever I try to read the newspaper, they come out and they MUST destroy it!" She opens the *Monadnock Shopper News* on the ground before her and, pretending to read it, mutters, sotto voce, "Gee, I'd like to read this article. . . . I really need that phone number there in that ad—I'll have to write it down. I sure hope nothing happens to the paper!"

A handful of hens immediately rush over. Elizabeth calls the rest. Rudy steps on the paper. Nikki, a Rhode Island Red, starts shredding the paper with her feet, holding it in her beak. "C'mere, Soot!" calls Elizabeth to a Black Sex-Link with a tall comb. The bird looks up, comes directly to Elizabeth, and starts pecking at the print. "Get that paper! Get that paper!" Elizabeth says encouragingly.

All the hens were raised on newspaper bedding, she explains, and they still love to scratch and shred it as they did as chicks. "Oh, look, an ad for Ocean State Job Lot," she says to the assembled group. "Sometimes your roost towels come from there."

"They like the Job Lot advertising section," she tells me, "because it's so colorful, I think." Goldie now arrives and starts shredding the paper with her feet. Peanut comes and pokes a hole in it with her beak.

Peanut, Jan, Nikki, and Soot all know their names. "The commands they know are impressive," says Elizabeth. "When I was carrying wood shavings for the floor of the coop and they would mob me, I would say to them, 'Back-back-back'—now all I have to do is say that and they stand out of the way." They will even back off, when she asks them, when they are massing at the opening to the pen or when Elizabeth is coming into their henhouse. They also respond to "in-in-in" and will rush back to the coop if they're near it. To call them from farther distances, Elizabeth gives the *chiva-chiva-chiva* whistle.

"If I'm sitting, I'll say, 'Coming up' and they'll jump onto my lap or my back. They also know the show of empty hands and 'all gone' and will walk away, realizing there is no more bread or other treats," she tells me.

Some people often attribute to their pets (and their children) intelligence and insight they don't have. Animal people are often accused of anthropomorphism, projecting human motives and emotions onto animals. But this is not the case with Elizabeth. She does not think of her Rangers as feathered people. She understands that they are different from us; they experience the world in ways we cannot imagine. They can see polarized light; they may hear in a different range; they can fly. They think and feel—but not always as we do. Right now, for instance, several Rangers, including Jan, are molting their feathers. Unlike my Ladies, whose molts have always been subtle affairs, molting Rangers sport ugly red bald patches and itchy-looking areas where new pinfeathers are coming in. When one molts, often the others will try to exclude her from the coop. Elizabeth has to intervene.

"There is a lack of pity in all birds," Elizabeth says thoughtfully as she feeds Peanut a bit of bread. "Chickens have no sympathy. People can't relate to it, so they don't like it. It's something we're not used to. But I like that about them," she says. "There's something very brave about them."

I'm deeply moved by Elizabeth's relationships with her chickens; I'm impressed by all the words, tricks, and games they know; I admire her astute observations. My affection for her and her Rangers grows with each day.

There's only one problem: our two flocks hate each other.

I didn't initially worry about how they would get along. After all, Elizabeth's Rangers were fenced, and thus, except for Rudy's daily escapes, confined by day to the roughly ten-foot-by-twelve-foot chicken yard she had erected outside their coop. My Ladies were never fenced, free to range over the property as they chose. Only during the hour or so Elizabeth spent with her hens in the afternoons, and during their time together on weekends, would the Rangers and the Ladies be loose at the same time. There would be little opportunity for the two flocks to fight. Because my Ladies had always accepted with alacrity new chicks added to their flock, I even hoped they might welcome the Rangers.

But I was wrong.

No fights ever broke out. No feathers flew. But after the Rangers moved in, the Ladies, to my great surprise, simply abandoned most of the territory the flock had held for twenty years.

They moved into Bobbie and Jarvis's yard.

I asked the Chicken Whisperer what she thought was happening. She admitted her Rangers were the aggressors. They were taunting my Ladies from behind the fence!

"Through the fence, they act like roosters, with their wings down to the ground like 'OHHHHH—I'm going to kick your tail!' They love doing that. They're really happy doing that. They'll leave their food to do that.

"I think chickens enjoy being in control and dominating others," she continued. Her beloved Peanut, the lowest in the pecking order since her surgery, is one of the worst aggressors: when one of my Ladies comes to investigate, Peanut raises her hackles and leaps to spar at the fence between them.

Now the Ladies never venture past the Rangers' fence. They visit the compost pile when the Rangers are enclosed. But when the Rangers are out, the Ladies take refuge next door.

I viewed the situation as a clash of cultures. Could this be possible? I asked Elizabeth.

"Every flock is quite different," Elizabeth agreed. She found this when she raised a second batch of chicks who came to complete the current flock. The older birds were far more domineering, exploratory, and brave. The difference between flocks was even more obvious comparing the Rangers with the Ladies.

The Rangers peck one another frequently and sometimes even peck Elizabeth—one pecked her by her eye, bruising and breaking the skin. Mine peck my palm when food is on it, but in my two decades of raising chickens, none has ever pecked my face. And though Elizabeth's hens are very affectionate, they do not like being stroked like mine do; unlike the Ladies, the Rangers never squat before her asking for a caress. Instead, they stand still before her, waiting to be picked up and kissed.

My Ladies have always been extremely calm and peaceful. The Rangers are more vocal, more aggressive, more domineering. The Rangers are rabble-rousers, always ready to rumble. My father-in-law would have said they were always making a tzimmes.

Everything the Rangers do is writ large. My hens are gentle, subtle; they are Ladies. The Rangers are drama queens.

Does this sound like anthropomorphism? Am I projecting onto chickens traits that belong to humans alone? How can it be that birds—a lineage that separated from that of the mammals more than three hundred million years ago—are as individual as people? How can these birds share with us intelligence, reasoning, foresight, memory—and in other ways slavishly obey blind instinct? Can creatures more closely related to lizards and crocodiles than to people actually have culture?

To me, it's clear: though none of my original Black Sex-Links of twenty years ago survives, their calm culture has persisted far longer than any individual's lifetime. And it has persisted through generations of unrelated chickens of different breeds.

All this while, they continually surprise me. Come one spring day, I saw, to my dismay, something I had never before witnessed.

My hens were crossing the road.

Why did the chickens cross the road?

I think it must have been to expand their territory, after the Rangers had taken over so much land they'd formerly considered their own. The neighbors across the street, along with their roaming Dalmatian, had moved, and their house sat empty, unraked leaves in their yard sheltering delicious bugs and their larvae. But visiting this new destination brought new dangers to the flock. The road—even a rural street with little traffic—is a dangerous place for hens. One day I found the black-and-white body of one of my Ladies smashed in the middle of the tarmac. A broken brown egg oozing a bright yellow yolk lay just beyond her tail. A racing logging truck, busily deforesting a lot for a new house much farther up the road, was almost certainly the culprit. I called the police, who spoke to the loggers. But there was nothing else they could do but offer condolences.

The neighborhood was changing. Elizabeth found a new love, and she and her Rangers made a new home; a new tenant moved in with us. Sadly, Jarvis got sick. He and Bobbie moved to a retirement community with nursing care in the next town. One of their three fine sons bought their house, and we were glad it remained in the family. But because they already had another house, most of the time nobody was home next door. The hens realized this, and I am sure they missed the Coffins. But the most important change for the flock was not the neighbors who moved out, but the neighbors who moved in—neighbors who did not live in houses.

After two centuries of persecution by farmers and fur traders, most of New Hampshire's native predators were nearly gone when we'd first moved to the state in the 1980s. The last wolf had been killed in the late 1800s. Fishers, cat-sized members of the weasel family, were trapped out from all but our northernmost counties by 1900. Bobcats, once bountied, were pronounced nearly gone by 1970. But our arrival in the state coincided with the beginning of a wave of restoration: with the return of the forests, with some legal protections, and with fewer human hunters and trappers haunting the woods, some of our state's most spectacular wild animals were staging a comeback. There were even some newcomers: the wolves did not return, but now there were coyotes, migrants from the West who had interbred with northern wolves. Between 1972 and 1980 they had spread across the state from the far north to the seacoast north of Boston. Today coyotes can be heard singing in every county in our state.

My husband used to joke, "If you want to see wildlife, get a flock of hens." Even in the early days, it was true: half a dozen times, over the course of the years, we had been alerted by alarm squawks to look out the window to see a fox running off with one of our hens in its mouth. ("Drop that!" my husband once yelled to a fox from an upstairs window. Incredibly, the fox did, and our hen was unharmed.) Once Jarvis had watched helplessly from his porch as a Cooper's hawk fell from the sky upon one of our Ladies like a lawn dart. And there was the skunk, and another time a mink, the neighbor's dog—and once, an ermine. The tiny, ferocious weasel in its snowy winter coat had slipped through a crack in our barn's stone foundation on Christmas Eve and decapitated one of our hens. Come Christmas morning, as I brought the flock their traditional holiday meal of fresh, warm popcorn, I was greeted with the sight of my slain friend—and next, with the sight of the perpetrator: clad in its angel-white coat, the ermine poked its head from a small hole in the wall and stared at me fearlessly with coal-black eyes, its face smeared with fresh crimson blood.

I could not be angry with these predators. Their return was the signal of a recovering ecosystem. We humans and our beloved pets are, after all, living on *their* land. Their kind had dwelt here long before people arrived on this continent. And as fearful as I was for the survival of our flock, every time I saw one of these native wild animals, I was thrilled. I would never hurt one of them.

By the late 2010s, we were seeing markedly more predators than ever before. One summer and fall, we regularly watched foxes trot down the street in broad daylight. One had a den under a neighbor's shed. At dusk we used to watch them from her porch. The vixen once returned from a hunting expedition bearing what looked, adorably, like a bent stick for the babies to play with. With binoculars we soon saw the "stick" had a yellow, scaly, four-toed foot on one end. It wasn't one of ours—that time, at least. But our flock was dwindling each year.

One day a bobcat seized one of our Ladies—again, during the daytime. The next afternoon, Howard and I were standing by the main floor of our barn, looking down upon some of her feathers that still remained, discussing the incident. "The bobcat was *right there*!" I said, pointing not ten feet from where we were standing, where we had caught just a glimpse—and at that very moment, the bobcat appeared again, so close we could clearly see the black top and white underside of the distinctive short tail. He never even looked at us, and strolled away nonchalantly as if he owned the place. Which, of course, he did.

The flock was then safe inside the coop. But we knew the bobcat would come back. Predators have excellent memories and are not apt to forget where they encountered a good meal. When I could not sit outside with them, watching them—which was most of the time, given my writing and frequent travels—my Ladies would be at risk. And for this reason, they were, to our mutual sorrow, increasingly confined.

We had a fence, but as protection, it was ineffective. Before Jarvis and Bobbie moved, Jarvis had hired a neighbor to build a fence connected to the coop. Four feet high and made of chicken wire, it was not really meant to keep our Ladies safe from danger, but to protect the Coffins' new Labrador retriever puppy. Unlike their previous Lab, this one delighted in eating chicken poop (many dogs apparently find this delicious), resulting in fire-hose diarrhea, usually loosed upon Bobbie's spotless kitchen. The fence did keep most poop separated from pup, but several of our hens quickly discovered they could fly over the fence, and did so daily.

Not even the hens who stayed inside the fence were safe. Most predators can dig under a fence. Hawks and owls can always attack from the sky. In fact, we often wondered if the fence actually endangered the hens it was meant to protect. Gretchen had told me gruesome tales of raccoons who, with their dexterous black hands, had grabbed hens through the chicken wire, and literally pureed the birds through the fence. If left free in the open, at least each hen had a chance to escape to hiding places they knew well among our eight acres of bushes, trees, gardens, and sheds. Inside a fence, they had fewer options if indeed an enemy got in; and because they were all confined in one space, any predator had an opportunity to kill not just one hen, but several—or all of them.

By the end of the summer of 2018, we were down to only three Dominiques. One afternoon, as I was about to head out for an event at a bookstore, I went to give them some vegetable peelings to entertain them while I was away and they were stuck inside. Two black-and-white bodies lay lifeless on the shavings. Both were missing their heads. Though we regularly patched small chinks in the barn's stone foundation, though we had assiduously dug up the dirt floor of the coop and lined it with small-mesh wire, there is almost no way to effectively exclude a weasel from a 150-year-old wood-and-stone barn.

The weasel, I knew, would soon return. If she stayed where she was, the lone survivor stood no chance. I could move her into the house . . . but then what? Chickens are flocking birds. They need company to feel safe, happy, and whole.

Should I get more birds? Raise more chicks in my office to start a new flock? It was certainly possible, given enough effort, time, and money, to design and build a new, super-safe coop within our barn and erect a new, secure, canopied fence. But under these conditions, no longer could they roam free while I gardened or hung the laundry. No more could they join us at the picnic table under the silver maple as we dined or lunched with friends. No longer would they rush to greet us as we emerged from the house or car. In fact, we would seldom even see them. I felt I owed my flock a chance to range free. But this was no longer an option.

These issues aside, I had to solve the immediate problem of what to do with my traumatized survivor. I gathered her up into my arms, put her in a box in the bathroom, and called our neighbor Julie Brown.

Julie, her husband, Phil, and their young kids, Laurel and Alden, keep a flock of about a dozen gentle, affectionate hens they call the Girls. They live less than a mile away from us, down a long, tree-lined driveway far from the street. I often cared for their hens when they were away. They have an ideal setup: a coop on stilts with a secure wooden floor. A spacious fenced and roofed area outside, protected with an electric wire. And the entire complex opens to their front yard, where the hens are free to range during the many hours most days that the kids are playing outside, or Julie is tending the family's numerous flower and vegetable gardens.

I brought my hen over before leaving for my event. Julie's gracious Girls accepted the survivor right away. She became best friends with a similarly colored Plymouth Rock. And there she stayed. I am welcome to visit anytime, and I do so often.

Last summer, the entire flock came to visit *us*. The Browns were away and I was caring for the hens at Julie and Phil's henhouse. One morning, when I came to let them out of the coop, I saw they were already out. Had I somehow failed to lock them up safe the night before? Had another neighbor let them loose that morning? I went around to the back of the coop and discovered how they got out: the entire back wall of the sturdy structure had been torn away by long, powerful claws. A bear had come to visit, and had carried off two hens. I was heartbroken—but relieved to see that my hen had once again survived disaster.

I did not have the skills or equipment to rebuild the coop—not even my husband could do this by nightfall, when the bear would surely be back. But what I did have was a perfectly good coop, far enough away from this one that the bear didn't know about it.

I knew exactly how to proceed. After quickly sweeping the cobwebs from our henhouse, adding fresh shavings, and setting up our old waterer and feeder, I fetched a pet carrier to transport our old hen, with her best friend, first. After I captured and moved the rest of the flock, two hens at a time, I let them settle in for a while before returning to the coop to sit among them that evening.

Of course, I was devastated about losing the two hens to the bear, and I hated that this had happened on my watch. But it could have been much worse. I sent Julie a photo of the damage, so that she and Phil could start planning the repairs. When they returned from their holiday, Phil would quickly rebuild the structure—and this time, Julie told me, they would encircle the entire complex, not just the fenced area, with electric wire, to deter another break-in from the bear.

But meanwhile, it was bliss to be able to stand among hens in our old coop again, surrounded by lilting chicken voices, enveloped in avian camaraderie and peace. Despite the recent tragedy, everyone seemed relaxed and at home. I credited my old Dominique for helping the rest of the flock feel comfortable and safe. I had no doubt she remembered her old home: for there she was, roosting on the same perch, in the exact same spot on that perch, where she had spent each night the first five years of her life—now, as then, with a trusted friend by her side.

At times like these, I like to remind myself how extraordinary it all is: chickens, like all birds, are really feathered dinosaurs. Unlike we mammals, their bones are hollow, their bodies are filled with air sacs; they are creatures made less of flesh than of air. And yet we share a fundamental talent: a need for companionship, a capacity for affection. These common creatures will never cease to dazzle me, with both our sameness and our differences. When I am with the flock, even when most of the birds are Julie's, and even when I am just visiting, they feel like family to me— and in a very real sense, they are.

SELECTED BIBLIOGRAPHY

Barber, Theodore Xenophon. *The Human Nature of Birds: A Scientific Discovery with Startling Implications.* New York: St. Martin's Press, 1993.

Cackle Hatchery Chicken catalogue: www.cacklehatchery .com.

Caughey, Melissa. *How to Speak Chicken: Why Your Chickens Do What They Do & Say What They Say.* North Adams, MA: Storey Publishing, 2017.

Danovich, Tove. *Under the Henfluence: Inside the World of Backyard Chickens and the People Who Love Them.* Evanston, IL: Agate Publishing, 2023.

Evans, C. S., and Linda Evans. "Chicken Food Calls Are Functionally Referential." *Animal Behaviour* 58, no. 2 (August 1999): 307–319.

Lorenz, Konrad. *On Aggression*. San Diego: Harcourt Brace, 1966.

Murray McMurray Chicken catalogue: www.mcmurray hatchery.com.

Rossier, Jay. *Living with Chickens: Everything You Need to Know to Raise Your Own Backyard Flock*. Guilford, CT: Lyons Press, 2004.

Skutch, Alexander F. *The Minds of Birds*. College Park: Texas A&M University Press, 1996.

Smith, C. L., and C. S. Evans. "Multimodal Signaling in Fowl, *Gallus gallus*." *Journal of Experimental Biology* 211, no. 13 (July 2008): 2052–2057.

Tate, Peter. *Flights of Fancy: Birds in Myth, Legend, and Superstition*. New York: Delacorte Press, 2007.

Wilson, D., and C. S. Evans. "Mating Success Increases Alarm-Calling Effort in Male Fowl, *Gallus gallus*." *Animal Behaviour* 76, no. 6 (December 2008): 2029–2035.

ABOUT THE PHOTOGRAPHER

Originally trained as an oil painter, Tianne Strombeck now creates photographic portraits of nature to promote conservation and wildlife education. Her painting background has given her an instinctive understanding of color and composition. She works to understand her subjects, to capture their essence, and to show how they interact with their environment and one another. Her work has appeared in numerous books, including Sy Montgomery's *The Soul of an Octopus*, *Condor Comeback*, and *The Hummingbirds' Gift*. To see additional examples of her work, from hummingbirds to jaguars, visit her galleries at https://www.tianimal.com.

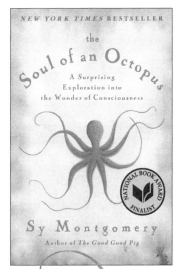
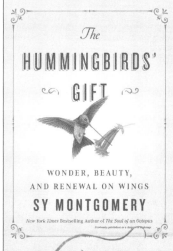